Groundworks®
Reasoning with Data and Probability

6

Carol Findell • Carole Greenes
Barbara Irvin • Jenny Tsankova

Acknowledgments

Dr. Carole Greenes
is a professor of Mathematics Education at Boston University. She has written and collaborated on more than 200 publications, and her work has brought teachers and students closer to the NCTM Standards. Dr. Greenes was recently inducted into the Massachusetts Mathematics Educators Hall of Fame.

Dr. Carol Findell
is a clinical associate professor of Education at Boston University. She has served as a mathematics educator for more than 30 years, during which time she has led writing teams for national mathematics competitions. Dr. Findell is a well-respected author and is a frequent speaker at national mathematics conferences.

Dr. Jenny Tsankova
is an assistant professor of Mathematics Education at Roger Williams University in Rhode Island. She has co-authored several articles as well as a primary-grade text. Dr. Tsankova serves on the Board of Directors of the Association of Teachers of Mathematics in Massachusetts.

Dr. Barbara Irvin
is an educational mathematics consultant. She has authored more than 30 activity resource books. Dr. Irvin is a former middle-school teacher, mathematics editor, and university professor.

Cover and interior illustrations by Debra Spina Dixon.

www.WrightGroup.com

Copyright ©2006 Wright Group/McGraw-Hill

All rights reserved. Except as permitted under the United States Copyright Act, no part of this publication may be reproduced or distributed in any form or by any means, or stored in a database or retrieval system, without the prior written permission from the publisher, unless otherwise indicated.

Permission is granted to reproduce the material contained on pages vii, 1–6, 9–14, 17–22, 25–30, 33–38, 41–46, 49–54, 57–62, 65–70, 73–78, 81–86, 89–94 and 96 on the condition that such material be reproduced only for classroom use; be provided to students, teachers, or families without charge; and be used solely in conjunction with *Groundworks: Reasoning with Data and Probability.*

Printed in the United States of America.

Send all inquiries to:
Wright Group/McGraw-Hill
P.O. Box 812960
Chicago, Illinois 60681

ISBN: 1-4045-3202-1

2 3 4 5 6 7 8 9 MAL 11 10 09 08 07 06

Contents

	Teacher Notes	iv
	Management Chart	vii
◆◇◇◇◇ **Interpret Data Displays**	Circle Story Fit the Facts	viii 8
◆◆◇◇◇ **Organize Data**	Average Distance Three Rings	16 24
◆◆◆◇◇ **Describe Data**	What is *N*? What Do You Mean? Number Switch	32 40 48
◆◆◆◆◇ **Ways to Count**	Side by Side Label Makers	56 64
◆◆◆◆◆ **Probability**	Blocks in the Box Marble Odds Mystery Number	72 80 88
	Certificate of Excellence	96

Why Teach *Reasoning with Data and Probability* to Your Students?

In their 1989 *Curriculum and Evaluation Standards for School Mathematics* and again in their 2000 *Principles and Standards for School Mathematics,* the National Council of Teachers of Mathematics identified data analysis and probability as one of the five major content strands of the mathematics curriculum, pre-kindergarten through grade 12. In school, children are continually presented with displays of data to interpret and the need to make decisions about the likelihood of events occurring. Outside of school, children play board, computer, and card games that involve interpreting and organizing information, identifying possible outcomes, and deciding which play will most likely lead to a win. Through these types of experiences, children develop an informal knowledge of and interest in many of the big ideas of data and probability that should be developed further.

Although NCTM and state framework groups recommend a developmental sequence of instruction in both data analysis and probability, mathematics programs tend to focus primarily on the interpretation of displays and less on the organization of data. Concepts of probability are often treated in a cursory fashion, or not explored at all, as is the case in most programs for primary-grade students. In general, the development of concepts of data and probability are not carefully sequenced within and across grades in elementary and middle school mathematics programs. *Groundworks: Reasoning with Data and Probability* focuses on five big ideas of data and probability, using interesting and challenging problems, and provides a sequence across grades that leads to greater understanding of data and probability concepts and skills.

Bibliography

Aspinwall, L and K. Shaw. "Enriching Students' Mathematical Intuitions with Probability Games and Tree Diagrams." *Mathematics Teaching in the Middle School* 6 (December 2000): pp. 214–220.

Dessart, D. and C. De Ridder. "Readers Write: Is Rock, Scissors, and Paper a Fair Game?" *Mathematics Teaching in the Middle School* 5 (September 1999): pp. 4–5.

Ewbank, W. and J. Ginther. "Probability on a Budget." *Mathematics Teaching in the Middle School* 7 (January 2002): pp. 280–283.

Irvin, B. *Up-to-Speed Math: Data Analysis and Reasoning.* Irvine, CA: Saddleback Publishing Co., 2002.

National Council of Teachers of Mathematics. *Navigating Through Data Analysis and Probability in Grades 6–8.* Reston, VA: NCTM, 2002.

National Council of Teachers of Mathematics. *Principles and Standards for School Mathematics.* Reston, VA: NCTM, 2000.

What Are the Five Big Ideas of *Reasoning with Data and Probability*?

Groundworks: Reasoning with Data and Probability for grades 1–7 provides challenging development of five big ideas of data and probability. Problems build on students' prior and current experiences and broaden and solidify their conceptual understanding of data and probability. The five big ideas are:

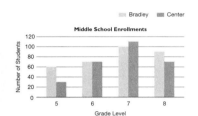

Interpret Data Displays
Students describe, compare, draw inferences, and predict from data presented in charts, frequency tables, pictographs, line plots, single and double bar graphs, circle graphs, stem-and-leaf diagrams, tree diagrams, and scatter plots. They learn to differentiate between discrete and continuous data. They also identify different representations of the same set of data.

Organize Data
Students learn how to organize information in displays in order to clarify relationships among the data. They sort and classify objects by a specified attribute and construct organized lists, Venn diagrams, tree diagrams, scatter plots, and line graphs. They use lists, tables, and tree diagrams to identify outcomes of experiments.

Describe Data
Students learn to describe sets of data using the mode, median, mean, range, and maximum and minimum values. They identify each of these measures in various types of data displays.

Ways to Count
Students develop understanding of three different ways to count when exploring probability. The *Fundamental Counting Principle* states that the number of outcomes of an event is the product of the number of outcomes for each part of the event. A *permutation* is an arrangement of a group of items in a particular order. A *combination* is a group of items in which the order of items is not important.

Probability
Students identify sample spaces and outcomes of compound events. They compute simple probabilities, and then compound probabilities. They explore the concept of odds, and the relationship of odds to probability. Students examine different games to determine their fairness.

What is in This Book?

This book contains:

- 12 blackline-master problem sets (72 pages of problems)
- solutions for all problems
- specific teaching suggestions and ideas for each problem set
- general teacher information

Problem Sets

Each problem set consists of eight pages. The first page presents teaching information, including goals identifying specific mathematical reasoning processes or skills, questions to ask students, and solutions to the first problem. The next six pages are all reproducible problem pages. The first problem is a teaching problem and is of moderate difficulty. The remaining five problems in the set range from easier to harder. Solutions to all problems are given on the eighth page of each problem set. For most problems, one solution method is shown; however, students may offer other valid methods. The mathematics required for the problems is in line with the generally approved mathematics curriculum for the grade level.

How to Use This Book

Because many of the problem types will be new to your students, you may want to have the entire class or groups of students work on the first problem in a set at the same time. You can use the questions that accompany the problem as the basis for a class discussion. As students work on the problem, help them with difficulties they may encounter. Students are frequently asked to explain their thinking. You may choose to do this orally with the whole class. After students have several experiences telling about their thinking and hearing the thinking of others, they are usually better able to write about their own thinking. Once students have completed the first problem in a set, you may want to assign the remaining problems for students to do on their own, in pairs in class, or for homework. If students have difficulty with the first problem in the set, you might do more of the problems with the whole class.

Although the big ideas and the families of problems within them are presented in a certain order, students need not complete them in this order. Students might work the problem sets based on the mathematical content of the problems and their alignment with your curriculum, or according to student interests or needs.

There is a Management Chart that you may duplicate for each student to keep in a portfolio. You may award the Certificate of Excellence upon the successful completion of the problem sets for each big idea.

Management Chart

Name _____ Class _____ Teacher _____

BIG IDEA	PROBLEM SET				DATE
◆◇◇◇◇ **Interpret Data Displays**	Circle Story	1	2	3	
		4	5	6	
	Fit the Facts	1	2	3	
		4	5	6	
◆◆◇◇◇ **Organize Data**	Average Distance	1	2	3	
		4	5	6	
	Three Rings	1	2	3	
		4	5	6	
◆◆◆◇◇ **Describe Data**	What is *N*?	1	2	3	
		4	5	6	
	What Do You Mean?	1	2	3	
		4	5	6	
	Number Switch	1	2	3	
		4	5	6	
◆◆◆◆◇ **Ways to Count**	Side by Side	1	2	3	
		4	5	6	
	Label Makers	1	2	3	
		4	5	6	
◆◆◆◆◆ **Probability**	Blocks in the Box	1	2	3	
		4	5	6	
	Marble Odds	1	2	3	
		4	5	6	
	Mystery Number	1	2	3	
		4	5	6	

◆◇◇◇◇
Interpret Data Displays

Circle Story

Goals
- Interpret circle graphs.
- Identify corresponding parts of a schedule and a circle graph.
- Compute percentages.

Notes
Suggest that students begin solving each problem by determining the total amount of time from the beginning to the end of the schedule. They can then use that amount of time to compute the number of minutes or hours represented by the percents on the circle graph.

Solutions to all problems in this set appear on page 7.

Circle Story 1

Questions to Ask
- What is the start time of the first activity on the schedule? (2:50 p.m.) The end time of the last activity? (9:30 p.m.)
- How long is it from the start of the first activity to the end of the last? (6 hours 40 minutes) How many minutes is that? (400 minutes) How do you know? ((6 × 60) + 40 = 400)
- Which activity took the greatest amount of time? (homework) Which section of the circle graph do you think shows the percent of time spent on homework? (Section a. 40%)
- How can you figure out if homework actually took 40% of the time? (First, determine the amount of time spent doing homework: from 3:50 p.m. to 6:30 p.m. is 160 minutes. Then find 40% of the total number of minutes: 40% of 400 minutes is 160 minutes. Finally, check that the two times match: 160 minutes = 160 minutes.)

Solutions
1. A Homework
   B Dinner
   C Walk home
   D Practice clarinet
   E Play outside
2. 70
3. 25

Reasoning with Data and Probability

Circle Story 1

Use Kevin's after-school schedule to identify the activities in the circle graph.

Activity	Start Time	End Time
Walk home	2:50 p.m.	3:10 p.m.
Play outside	3:10 p.m.	3:50 p.m.
Homework	3:50 p.m.	6:30 p.m.
Dinner	6:30 p.m.	7:30 p.m.
Practice clarinet	7:30 p.m.	9:30 p.m.
Lights out	9:30 p.m.	

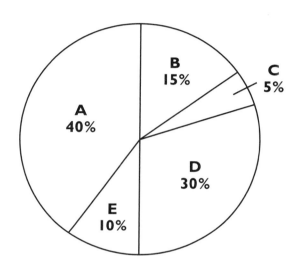

1. Write the activity for each section in the circle graph.

   Section A _____

   Section B _____

   Section C _____

   Section D _____

   Section E _____

2. What percent of Kevin's after-school time was spent doing homework and practicing the clarinet?

   _____

3. Suppose Kevin spent half as much time practicing the clarinet and spent the extra time playing outside. What percent of his time would he then spend playing outside?

   _____

Reasoning with Data and Probability

GROUNDWORKS 1

Circle Story 2

Use Catie's after-school schedule to identify the activities in the circle graph.

Activity	Start Time	End Time
Soccer practice	4:00 p.m.	5:30 p.m.
Practice violin	5:30 p.m.	6:00 p.m.
Dinner	6:00 p.m.	7:00 p.m.
Homework	7:00 p.m.	9:00 p.m.
Lights out	9:00 p.m.	

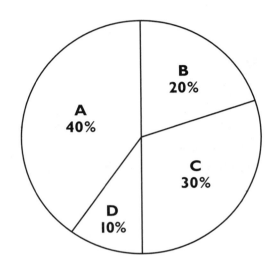

1. Write the activity for each section in the circle graph.

    Section A _____

    Section B _____

    Section C _____

    Section D _____

2. What percent of Catie's after-school time was spent doing homework and practicing the violin?

    _____

3. Suppose Catie spent half as much time doing homework and spent the extra time practicing violin. What percent of her time would she then spend practicing violin?

    _____

Circle Story 3

Use Alexa's after-school schedule to identify the activities in the circle graph.

Activity	Start Time	End Time
Walk dog	4:00 p.m.	4:30 p.m.
Homework	4:30 p.m.	5:30 p.m.
Read riddle book	5:30 p.m.	5:45 p.m.
Dinner and wash dishes	5:45 p.m.	7:00 p.m.
Knit sweater	7:00 p.m.	9:00 p.m.
Lights out	9:00 p.m.	

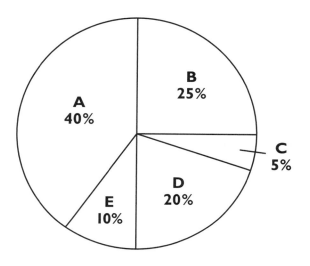

1. Write the activity for each section in the circle graph.

   Section A _____

   Section B _____

   Section C _____

   Section D _____

   Section E _____

2. What percent of Alexa's after-school time was spent doing homework and knitting?

   _____

3. Suppose Alexa finished knitting the sweater in one hour and spent the extra time reading the riddle book. What percent of her time would she then spend reading the riddle book?

   _____

Name _____

Interpret Data Displays

Circle Story ④

Use Mariko's after-school schedule to identify the activities in the circle graph.

Activity	Start Time	End Time
Piano lesson	4:00 p.m.	4:45 p.m.
Gymnastics	4:45 p.m.	6:00 p.m.
Ride bike	6:00 p.m.	6:30 p.m.
Dinner	6:30 p.m.	7:30 p.m.
Homework	7:30 p.m.	9:00 p.m.
Lights out	9:00 p.m.	

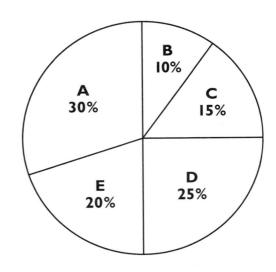

1. Write the activity for each section in the circle graph.

    Section A _____

    Section B _____

    Section C _____

    Section D _____

    Section E _____

2. What percent of Mariko's after-school time was spent doing gymnastics and riding her bike?

    _____

3. Suppose Mariko did not have gymnastics or a piano lesson. Instead, she practiced the piano for one hour and spent the rest of the extra time reading a mystery book. What percent of her time would she then spend reading the mystery book?

    _____

Circle Story 5

Use Leland's after-school schedule to identify the activities in the circle graph.

Activity	Start Time	End Time
Bus ride home	2:50 p.m.	3:10 p.m.
Walk dog	3:10 p.m.	3:50 p.m.
Practice piano	3:50 p.m.	4:50 p.m.
Homework	4:50 p.m.	6:50 p.m.
Dinner	6:50 p.m.	7:30 p.m.
Read	7:30 p.m.	9:30 p.m.
Lights out	9:30 p.m.	

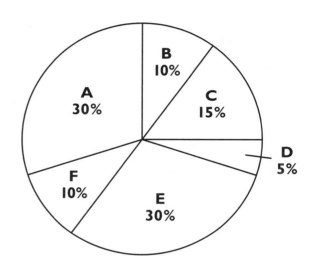

1. Write the activity for each section in the circle graph.

    Section A _____   Section B _____

    Section C _____   Section D _____

    Section E _____   Section F _____

2. Suppose Leland spent 40 minutes of his reading time on homework. What percent of his time would he then spend on reading?

    _____

3. Suppose Leland walked home instead of taking the bus, and the walk took twice as long as the bus ride. Leland took off the extra walking time from his piano practice time. What percent of his time would he then spend on practicing the piano?

    _____

Circle Story 6

Use Bob's after-school schedule to identify the activities in the circle graph.

Activity	Start Time	End Time
Bike home	2:50 p.m.	3:00 p.m.
Eat snack	3:00 p.m.	3:20 p.m.
Homework	3:20 p.m.	5:20 p.m.
Practice guitar	5:20 p.m.	6:00 p.m.
Dinner	6:00 p.m.	7:00 p.m.
Basketball game	7:00 p.m.	9:30 p.m.
Lights out	9:30 p.m.	

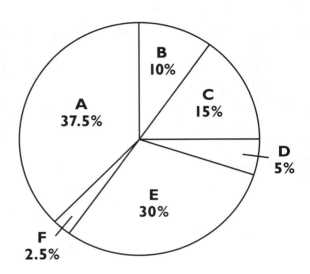

1. Write the activity for each section in the circle graph.

   Section A _____   Section B _____

   Section C _____   Section D _____

   Section E _____   Section F _____

2. What percent of Bob's time was spent biking home, doing homework, and practicing the guitar?

   _____

3. Suppose Bob played basketball for $\frac{5}{6}$ of an hour and spent the extra time reading a book. What percent of his time would he then spend reading a book and practicing the guitar?

   _____

Solutions

Circle Story 1

1. A Homework
   B Dinner
   C Walk home
   D Practice clarinet
   E Play outside
2. 70
3. 25

Circle Story 2

1. A Homework
   B Dinner
   C Soccer
   D Practice violin
2. 50
3. 30

Circle Story 3

1. A Knit sweater
   B Dinner and dishes
   C Read book
   D Homework
   E Walk dog
2. 60
3. 25

Circle Story 4

1. A Homework
   B Ride bike
   C Piano lesson
   D Gymnastics
   E Dinner
2. 35
3. 20

Circle Story 5

1. A Homework or Reading
   B Walk dog or Dinner
   C Practice piano
   D Bus ride home
   E Reading or Homework
   F Dinner or Walk dog
2. 20
3. 10

Circle Story 6

1. A Basketball game
   B Practice guitar
   C Dinner
   D Eat snack
   E Homework
   F Bike home
2. 42.5
3. 35

♦◊◊◊◊
Interpret Data Displays

Fit the Facts

Goals
- Interpret double bar graphs.
- Find the mean and median of a set of data.
- Compute percentages.

Notes
Before students fill in the blanks, suggest that they first identify the quantity represented by each bar. Point out that the blanks do not have to be completed in order.
Solutions to all problems in this set appear on page 15.

Fit the Facts I

Questions to Ask
- What do the lighter and darker bars represent? (The lighter bars represent populations in 1960. The darker bars represent populations predicted for 2010.)
- What was the population of Darnell in 1960? (40,000 people)
- What is the predicted population for Hart in 2010? (60,000 people)
- In 1960, which city had 50% of the population of another city? (Central City had 50% of the population of Hart.) How did you figure it out? (Central City's population was 25,000 people and Hart's population was 50,000 people; 50%, or $\frac{1}{2}$, of 50,000 is 25,000.)
- What is the mean of the 1960 populations of Darnell and Hart? (45,000 people) How did you figure it out? (40,000 + 50,000 = 90,000, and 90,000 ÷ 2 = 45,000)

Solutions
1. Central City
2. Hart
3.–5. Darnell, Elksville, Forest Town
6. Forest Town
7. Darnell
8. 40,000
9. 47,000

GROUNDWORKS

Reasoning with Data and Probability

Fit the Facts ◆1

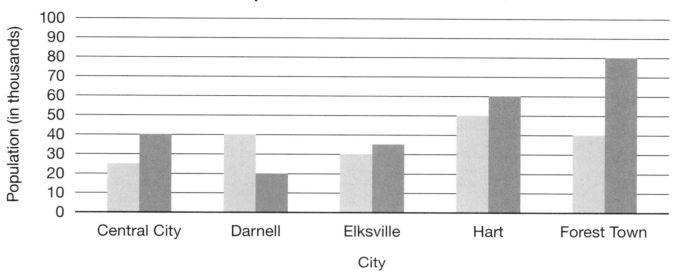

Use the information in the bar graph to fill in the blanks.

In 1960, _____(1)_____ had 50% of the population of _____(2)_____.

The average population of _____(3)_____, _____(4)_____, and _____(5)_____ in 1960 was about 37,000 people. It is predicted that by 2010, the population of _____(6)_____ will increase by 100% and the population of _____(7)_____ will decrease by 50%. In 2010, the median population of the five cities will be _____(8)_____ people, and the mean population will be _____(9)_____ people.

Fit the Facts 2

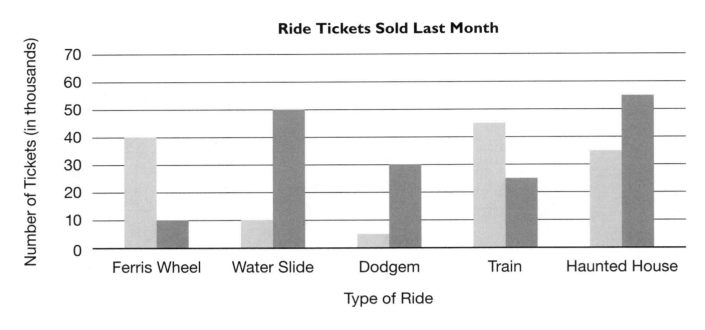

Use the information in the bar graph to fill in the blanks.

The median number of child tickets sold last month was _____(1)_____; the median

number of adult tickets sold was _____(2)_____. The mean number of child tickets

sold was _____(3)_____; the mean number of adult tickets sold was

_____(4)_____. Six times as many child as adult tickets were sold for the

_____(5)_____ ride. One-fourth as many child as adult tickets were sold for the

_____(7)_____ ride. The most popular ride was the _____(6)_____. The least

popular ride was the _____(8)_____.

Name _____

Interpret Data Displays

Fit the Facts ⟨3⟩

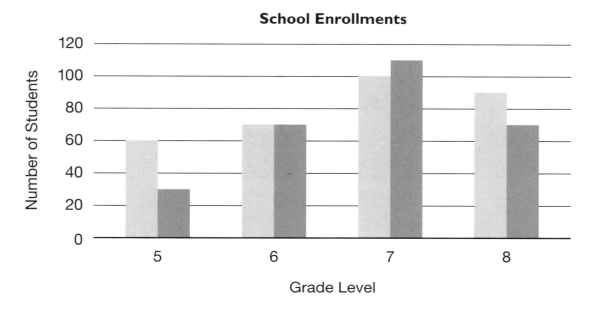

Use the information in the bar graph to fill in the blanks.

In Grade _____, the number of students in _____ School is
         (1)   (2)

100% greater than the number of students in _____ School.
                                                      (3)

In Grade _____, the number of students in _____ School is
         (4)   (5)

10% greater than the number of students in _____ School. The mean
                                               (6)

number of students in each grade in _____ School is 70. The mean number
                                                   (7)

of students in each grade in _____ School is 80.
                             (8)

Name _____

Interpret Data Displays

Fit the Facts ◆4

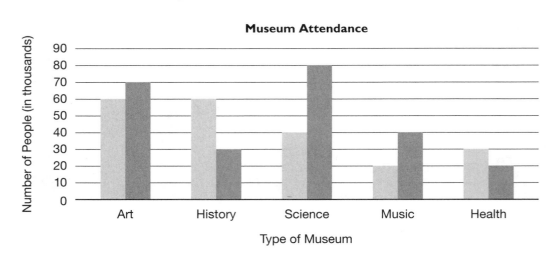

Use the information in the bar graph to fill in the blanks.

In 2000, attendance at the _____(1)_____ museum was 50% of the attendance at

both the _____(2)_____ museum and the _____(3)_____ museum. From 2000 to

2005, attendance at the _____(4)_____ museum decreased by 50%, and

attendance at both the _____(5)_____ museum and the _____(6)_____ museum

increased by 100%. In 2000, the average number of people who attended the

_____(7)_____, _____(8)_____, _____(9)_____, and

_____(10)_____ museums was 45,000. In 2000, the median number of people who

attended a museum was _____(11)_____. In 2005, the mean number of people who

attended a museum was _____(12)_____.

12 GROUNDWORKS

Reasoning with Data and Probability

Name _____

Interpret Data Displays

Fit the Facts ⟨5⟩

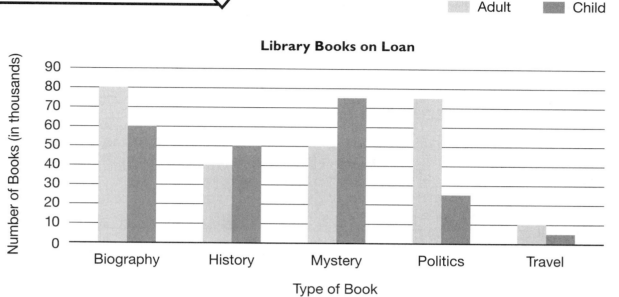

Use the information in the bar graph to fill in the blanks.

The number of adult _____(1)_____ books taken out was two-thirds the number of child books taken out on the same topic. The number of adult _____(2)_____ books taken out was 100% greater than the number of child books taken out on the same topic. The median number of child books taken out was _____(3)_____, and the median number of adult books taken out was _____(4)_____. The number of child books taken out ranged from _____(5)_____ to _____(6)_____, with a mean of _____(7)_____. The number of adult books taken out ranged from _____(8)_____ to _____(9)_____, with a mean of _____(10)_____.

Reasoning with Data and Probability

GROUNDWORKS ⟨13⟩

Fit the Facts ⟨6⟩

Interpret Data Displays

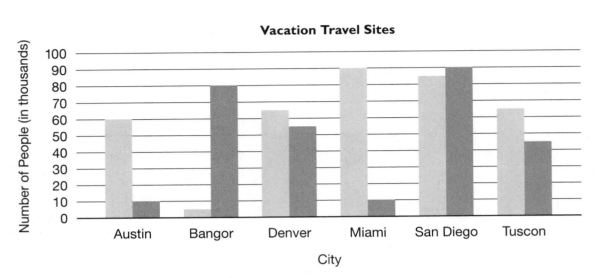

Use the information in the bar graph to fill in the blanks.

The median number of travelers to the six sites in January was _____(1)_____ ; in

June, the median number of travelers was _____(2)_____ . The most popular site in

January was _____(3)_____ , which was 18 times as popular as _____(4)_____

during the same month. In January, the number of travelers to _____(5)_____ was

50% greater than the number of travelers to _____(6)_____ . The average number of

travelers to _____(7)_____ , _____(8)_____ , and _____(9)_____ in June was

25,000. The range of the numbers of travelers in January was _____(10)_____ and in

June was _____(11)_____ .

Solutions

Fit the Facts 1

1. Central City
2. Hart
3.–5. Darnell, Elksville, Forest Town
6. Forest Town
7. Darnell
8. 40,000
9. 47,000

Fit the Facts 2

1. 30,000
2. 35,000
3. 34,000
4. 27,000
5. Dodgem
6. Ferris Wheel
7. Haunted House
8. Dodgem

Fit the Facts 3

1. 5
2. Bradley
3. Center
4. 7
5. Center
6. Bradley
7. Center
8. Bradley

Fit the Facts 4

1. health
2. art or history
3. history or art
4. history
5. music or science
6. science or music
7.–10. art, history, science, music
11. 40,000
12. 48,000

Fit the Facts 5

1. mystery
2. travel
3. 50,000
4. 50,000
5. 5,000
6. 75,000
7. 43,000
8. 10,000
9. 80,000
10. 51,000

Fit the Facts 6

1. 65,000
2. 50,000
3. Miami
4. Bangor
5. Miami
6. Austin
7.–9. Austin, Denver, Miami
10. 85,000
11. 80,000

◆◆◇◇◇
Organize Data

Average Distance

Goals
- Construct a scatter plot from a table of values.
- Identify the mean from two different representations of the same set of data.
- Estimate the mean from values on a scatter plot.
- Compute the mean and range from a set of data.

Notes
Explain to students that they can estimate the average distance by first imagining a horizontal line halfway between the highest and lowest points on the graph. They can then adjust their estimates by comparing the distances of other pairs of points above and below that line.

Solutions to all problems in this set appear on page 23.

Average Distance 1

Questions to Ask
- What is the least number of miles traveled for the 6 days? (250 miles)
- What is the greatest number of miles traveled? (350 miles)
- What is the mean of the greatest and least distances traveled? (300 miles) How do you know? ((350 + 250) ÷ 2 = 300)
- Suppose you were to draw a horizontal line on the graph halfway between the highest and lowest points. Would the second highest and second lowest points be about the same distance from the line? (yes) Would there be other pairs of points equidistant from the line? (yes) How many points would be above the line? (3) How many points would be below the line? (3)

Solutions

1. and 2.

3. Possible answer: Halfway between the greatest and least distances traveled is 300 miles.

4. 300 miles

GROUNDWORKS

Reasoning with Data and Probability

Organize Data

Average Distance 1

The Jones family recorded the number of miles traveled each day on their 6-day road trip.

1. Use the values in the table to construct a scatter plot.

Miles Traveled

Day	Distance (in mi)
1	350
2	310
3	250
4	290
5	275
6	325

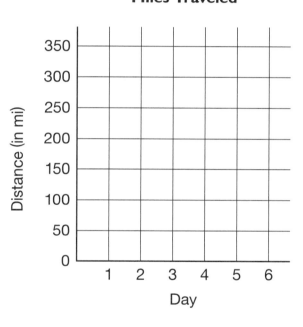

2. Estimate the average number of miles the Jones family traveled each day. Then draw a horizontal line on the graph to represent it.

3. Explain how you decided where to draw the line. _____

_____

4. What is the average distance the Jones family traveled each day?

_____

Reasoning with Data and Probability

GROUNDWORKS 17

Average Distance 2

Jackson recorded the number of miles traveled each day on his 5-day bicycle race.

1. Use the values in the table to construct a scatter plot.

Miles Traveled

Day	Distance (in mi)
1	90
2	100
3	120
4	80
5	110

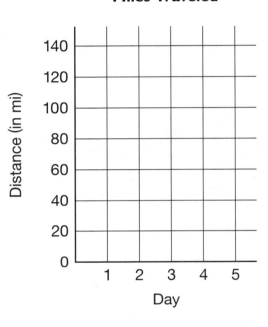

Miles Traveled

2. Estimate the average number of miles Jackson traveled each day. Then draw a horizontal line on the graph to represent it.

3. Explain how you decided where to draw the line. _____

4. What is the average distance Jackson traveled each day?

Average Distance ◇3

Ashley recorded the number of miles traveled each day on her 5-day running race.

1. Use the values in the table to construct a scatter plot.

Miles Traveled

Day	Distance (in mi)
1	22
2	20
3	28
4	25
5	30

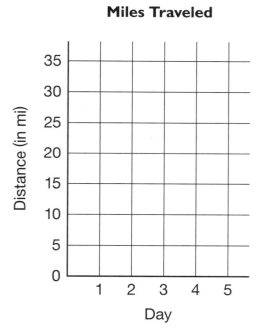

2. Estimate the average number of miles Ashley traveled each day. Then draw a horizontal line on the graph to represent it.

3. Explain how you decided where to draw the line. _____

_____

4. What is the average distance Ashley traveled each day?

_____

Name _____

Organize Data

Average Distance ◆4◆

Bob recorded the number of miles traveled each day on his 6-day hiking trip.

1. Use the values in the table to construct a scatter plot.

Miles Traveled

Day	Distance (in mi)
1	30
2	35
3	24
4	31
5	24
6	36

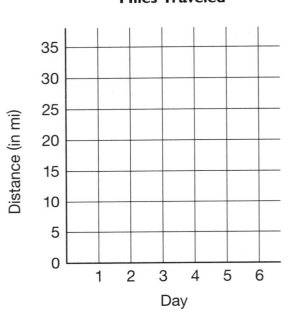

2. Estimate the average number of miles Bob traveled each day. Then draw a horizontal line on the graph to represent it.

3. Explain how you decided where to draw the line. _____

   _____

4. What is the average distance Bob traveled each day?

   _____

20 GROUNDWORKS   Reasoning with Data and Probability

Name _____

Organize Data

Average Distance ⟨5⟩

Maria recorded the number of miles traveled each day on her 6-day canoe trip.

1. Use the values in the table to construct a scatter plot.

Miles Traveled

Day	Distance (in mi)
1	10
2	12
3	20
4	16
5	18
6	12

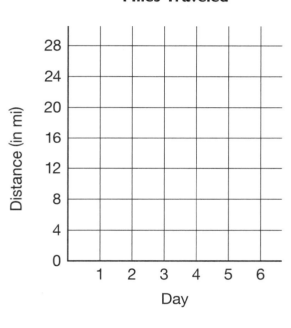

Miles Traveled

2. Estimate the average number of miles Maria traveled each day. Then draw a horizontal line on the graph to represent it.

3. Explain how you decided where to draw the line. _____

_____

4. What is the average distance Maria traveled each day?

_____

Name _____

Organize Data

Average Distance ◇6◇

Mr. Ramirez recorded the number of miles traveled each day on his 6-day car race.

1. Use the values in the table to construct a scatter plot.

Miles Traveled

Day	Distance (in mi)
1	650
2	600
3	625
4	700
5	550
6	575

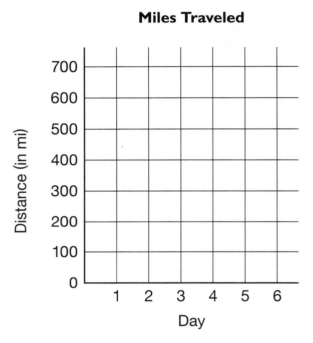

Miles Traveled

2. Estimate the average number of miles Mr. Ramirez traveled each day. Then draw a horizontal line on the graph to represent it.

3. Explain how you decided where to draw the line. _____

_____

4. What is the average distance Mr. Ramirez traveled each day?

_____

22 GROUNDWORKS   Reasoning with Data and Probability

Solutions

Average Distance

1. and 2.
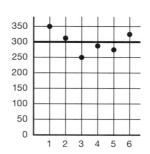

3. Possible answer: Halfway between the greatest and least distances traveled is 300 miles.

4. 300 miles

Average Distance

1. and 2.
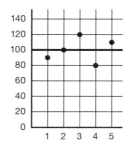

3. Possible answer: Halfway between the greatest and least distances traveled is 100 miles.

4. 100 miles

Average Distance

1. and 2.
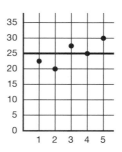

3. Possible answer: Halfway between the greatest and least distances traveled is 25 miles.

4. 25 miles

Average Distance

1. and 2.
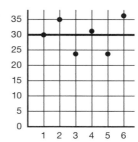

3. Possible answer: Halfway between the greatest and least distances traveled is 30 miles.

4. 30 miles

Average Distance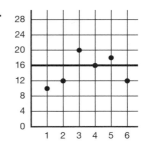

1. and 2.

3. Possible answer: Halfway between the greatest and least distances traveled is 15 miles.

4. 14.7 miles

Average Distance

1. and 2.
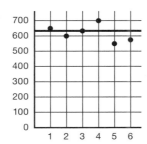

3. Possible answer: Halfway between the greatest and least distances traveled is 625 miles.

4. 616.7 miles

Organize Data

Three Rings

Goals
- Organize and interpret survey data using three-ring Venn diagrams.
- Use logical reasoning to solve problems.

Notes
Suggest to students that they record the total number of responses for each ring category next to the category name. Once they have numbers in all of the regions, they can add the numbers in the rings and check the sums with the category totals. Encourage students to begin with the region shared by all three rings and then consider regions that are shared by pairs of rings.

Solutions to all problems in this set appear on page 31.

Three Rings I

Questions to Ask
- How many people said all three types of books? (15) In which region will you write 15? (Region E)
- How many people said mystery books? (54) Which regions give a total of 54 people who like mystery books? (Regions D, E, F, and G)
- What does Region C represent? (the number of people who like history books but do not like nature or mystery books)
- What is the sum of the numbers of people in Regions A through H? (125)

Solutions

1.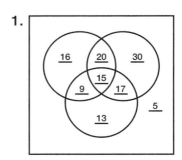

2. 16
3. 13
4. 20
5. 5

GROUNDWORKS   Reasoning with Data and Probability

Organize Data

Three Rings ①

Samantha asked 125 people, "What is your favorite type of book?"

1. Use the responses shown to fill in the number of people in each region of the Venn diagram.

125 People Responded
15 said all three types of books
60 said nature books
54 said mystery books
82 said history books
35 said nature and history books
24 said nature and mystery books
32 said mystery and history books

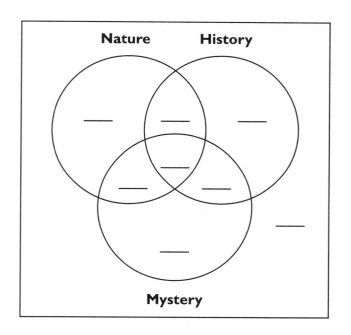

2. How many people said only nature books? _____

3. How many people said only mystery books? _____

4. How many people said nature and history but not mystery books? _____

5. How many people did not like any of the three types of books? _____

Three Rings ②

Alan asked 64 people, "What is your favorite type of museum?"

1. Use the responses shown to fill in the number of people in each region of the Venn diagram.

64 People Responded
2 said all three types of museums
6 said art and science museums
7 said art and natural history museums
10 said natural history and science museums
15 said only art museums
20 said only natural history museums
9 said only science museums

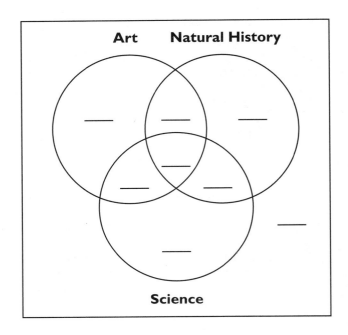

2. How many people said art and natural history but not science museums? _____

3. How many people said natural history and science but not art museums? _____

4. How many people said art and science but not natural history museums? _____

5. How many people did not like any of the three types of museums? _____

Three Rings 3

Jennifer asked 50 people, "What is your favorite type of music?"

1. Use the responses shown to fill in the number of people in each region of the Venn diagram.

50 People Responded
4 said all three types of music
10 said jazz and classical
12 said jazz and rock
7 said rock and classical
20 said jazz
24 said rock
23 said classical

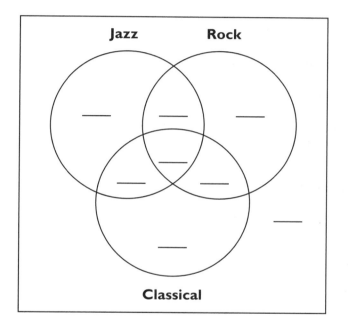

2. How many people said only jazz? _____

3. How many people said only rock? _____

4. How many people said rock and classical but not jazz? _____

5. How many people did not like any of the three types of music? _____

Three Rings ◆4

Luis asked 100 people, "What language did you study in school?"

1. Use the responses shown to fill in the number of people in each region of the Venn diagram.

100 People Responded
5 studied all three languages
20 studied French and Spanish
15 studied German and Spanish
11 studied French and German
50 studied Spanish
51 studied German
38 studied French
2 did not study any of the three languages

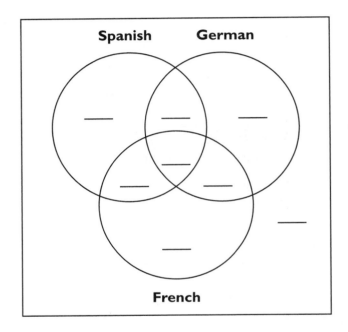

2. How many people studied only Spanish? _____

3. How many people studied only German? _____

4. How many people studied only French? _____

5. How many people studied Spanish or German but not both? _____

Three Rings ⟨5⟩

Susan asked some people, "What country would you like to visit?"

1. Use the responses shown to fill in the number of people in each region of the Venn diagram.

? People Responded
90 said Italy
70 said England
100 said Australia
30 said all three countries
40 said England and Australia
50 said Italy and Australia
45 said Italy and England
15 did not like any of the three countries

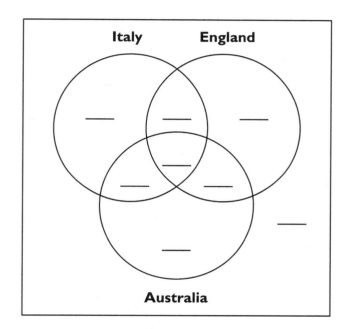

2. How many people were surveyed? _____

3. How many people said England and Australia but not Italy? _____

4. How many people said only Italy? _____

5. How many people said only Australia? _____

Three Rings ⟨6⟩

Julian asked some people, "What is your favorite type of juice?"

1. Use the responses shown to fill in the number of people in each region of the Venn diagram.

? People Responded
70 said orange juice
60 said apple juice
40 said grape juice
24 said apple and orange juices
16 said apple and grape juices
15 said orange and grape juices
14 said orange and apple juices but not grape juice
10 did not like any of the three juices

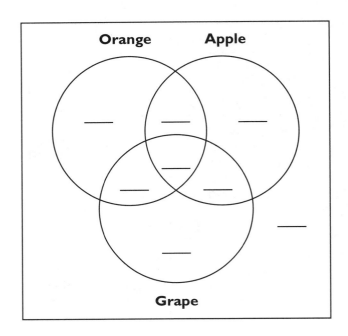

2. How many people were surveyed? _____

3. How many people said apple and grape juices but not orange juice? _____

4. How many people said only orange juice? _____

5. How many people said only grape juice? _____

Solutions

Three Rings

1.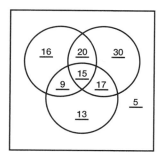

2. 16
3. 13
4. 20
5. 5

Three Rings 2

1.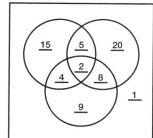

2. 5
3. 8
4. 4
5. 1

Three Rings 3

1.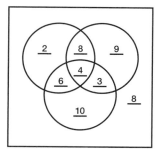

2. 2
3. 9
4. 3
5. 8

Three Rings 4

1.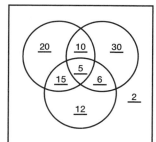

2. 20
3. 30
4. 12
5. 71

Three Rings 5

1.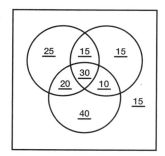

2. 170
3. 10
4. 25
5. 40

Three Rings 6

1.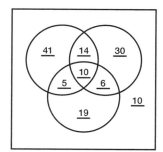

2. 135
3. 6
4. 41
5. 19

Describe Data

What is N?

Goals
- Identify the mean and the median of a set of data.
- Use logical reasoning to solve problems.

Notes
Prior to doing this problem set, review the concepts of mean and median and how to find them. To help identify candidates for N, suggest that students look at multiples of the number of numbers in each set and then compare that number with the sum of the values.

Solutions to all problems in this set appear on page 39.

What Is N? 1

Questions to Ask
- What type of number is N? (a counting number) Can N be zero? (no) Why not? (Zero is not a counting number.)
- What type of numbers are the mean and median of the set of numbers? (counting numbers) Can the mean or median be zero? (no) Why not? (Zero is not a counting number.)
- How many numbers are in the set? (5)
- What number must the sum of the numbers be divisible by? (5)
- How can you use this information to help you solve the problem? (Look for multiples of 5 that could be the sum of the five numbers.)
- Could 30 be the sum of the numbers? (no) Why not? (If the sum is 30, then N = 2. If N = 2, then the median is 4 and the mean is 6. Since the median and the mean are not the same, the sum cannot be 30.)

Solutions
1. 7
2. Possible answer: The sum of the given numbers is 28. Multiples of 5 more than but close to 28 are 30, which gives a value of 2 for N, and 35, which gives a value of 7 for N. If N is 2, the median is not equal to the mean. If N is 7, the median and the mean are both 7.

What is N? 1

N is a counting number.
The mean and the median are also counting numbers.

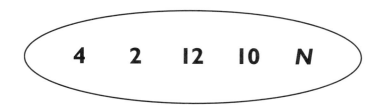

1. For what value of N would the mean and the median be

   the same? _____

2. How did you figure it out? _____

   _____

   _____

   _____

What is N? ②

N is a counting number.
The mean and the median are also counting numbers.

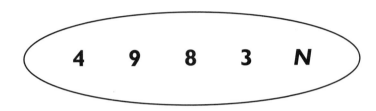

1. For what value of *N* would the mean and the median be

   the same? _____

2. How did you figure it out? _____

   _____

   _____

   _____

What is N? ⟨3⟩

N is a counting number.
The mean and the median are also counting numbers.

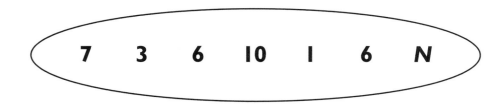

1. For what value of N would the mean and the median be

   the same? _____

2. How did you figure it out? _____

   _____

   _____

What is N? ◇4◇

Name _____

Describe Data

N is a counting number.
The mean and the median are also counting numbers.

7   6   3   9   3   4   4   6   N

1. For what value of N would the mean and the median be

   the same? _____

2. How did you figure it out? _____

   _____

   _____

   _____

Describe Data

What is N? ⟨5⟩

K and N are counting numbers.
The mean and the median are also counting numbers.

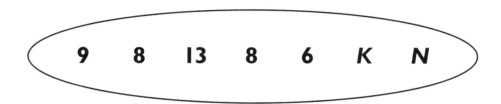

1. For what values of K and N would the mean and the median be

   the same? _____ and _____

2. How did you figure it out? _____

   _____

   _____

   _____

What is N? ⟨6⟩

K and *N* are counting numbers.
The mean and the median are also counting numbers.

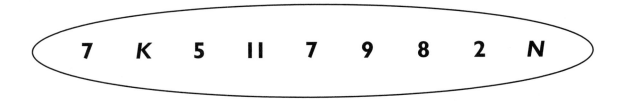

1. For what values of *K* and *N* would the mean and the median be

    the same? _____ and _____

2. How did you figure it out? _____

   _____

   _____

   _____

Solutions

What is N? 1

1. 7
2. Possible answer: The sum of the given numbers is 28. Multiples of 5 more than but close to 28 are 30, which gives a value of 2 for N, and 35, which gives a value of 7 for N. If N is 2, the median is not equal to the mean. If N is 7, the median and the mean are both 7.

What is N? 2

1. 6
2. Possible answer: The sum of the given numbers is 24. Multiples of 5 more than but close to 24 are 25, which gives a value of 1 for N, and 30, which gives a value of 6 for N. If N is 1, the median is not equal to the mean. If N is 6, the median and the mean are both 6.

What is N? 3

1. 9
2. Possible answer: The sum of the given numbers is 33. Multiples of 7 more than but close to 33 are 35, which gives a value of 2 for N, and 42, which gives a value of 9 for N. If N is 2, the median is not equal to the mean. If N is 9, the median and the mean are both 6.

What is N? 4

1. 12
2. Possible answer: The sum of the given numbers is 42. Multiples of 9 more than but close to 42 are 45, which gives a value of 3 for N, and 54, which gives a value of 12 for N. If N is 3, the mean is 5 but there is no 5 in the set for the median. If N is 12, the median and the mean are both 6.

What is N? 5

1. 1, 11 or 2, 10 or 3, 9 or 4, 8 or 5, 7 or 6, 6
2. Possible answer: The sum of the given numbers is 44. Multiples of 7 greater than but close to 44 are 49, which gives a value of 5 for K + N, and 56, which gives a value of 12 for K + N. If K + N is 5, the mean is 7 but there is no 7 in the set for the median. If K + N is 12, the median and the mean are both 8. So K and N can be 1 and 11 or 2 and 10 or 3 and 9 or 4 and 8 or 5 and 7 or 6 and 6.

What is N? 6

1. 1, 13 or 2, 12 or 3, 11 or 4, 10 or 5, 9 or 6, 8 or 7, 7
2. Possible answer: The sum of the given numbers is 49. Multiples of 9 greater than but close to 49 are 54, which gives a value of 5 for K + N, and 63, which gives a value of 14 for K + N. If K + N is 5, the mean is 6 but there is no 6 in the set for the median. If K + N is 14, the median and the mean are both 7. So K and N can be 1 and 13 or 2 and 12 or 3 and 11 or 4 and 10 or 5 and 9 or 6 and 8 or 7 and 7.

Describe Data

What Do You Mean?

Goals
- Compute and interpret the mean of a set of data.
- Given the mean of a set of numbers, find the greatest or least possible number in the set.

Notes
Prior to doing this problem set, review the concept of mean and how to compute it. Encourage students to think of numbers that are not close to the mean in order to find the least or greatest number in a set of numbers. For example, three numbers with a mean of 20 are 1, 2, and 57 since 1 + 2 + 57 = 60 and 60 ÷ 3 = 20. Remind students that zero is not a counting number.

Solutions to all problems in this set appear on page 47.

What Do You Mean? 1

Questions to Ask
- How many numbers is Elena thinking of? (6)
- Are some of her numbers the same? (No; they are all different.)
- What is the mean of the six numbers? (9)
- What is the least number possible in this set? (1) How do you know? (Elena says that the counting numbers she is thinking of are 100 or less.)
- What is the sum of Elena's numbers? (54) How do you know? (Since the mean of the six numbers is 9, their sum is 6 × 9, or 54.)

Solutions
1. 39
2. Possible answer: Since the mean of the six numbers is 9, their sum is 6 × 9, or 54. To get the greatest number, first find the sum of the five least counting numbers and then subtract that sum from 54: 1 + 2 + 3 + 4 + 5 = 15, and 54 − 15 = 39. So 39 is the greatest possible number.

Name _____

Describe Data

What Do You Mean?

"I am thinking of six counting numbers that are 100 or less. All of the numbers are different. The mean of my six numbers is 9."

Elena

1. What is the greatest number possible for one of Elena's numbers? _____

2. How did you figure it out? _____

_____

_____

_____

Reasoning with Data and Probability

GROUNDWORKS 41

What Do You Mean?

Nathan

"I am thinking of four counting numbers that are 100 or less. All of the numbers are different. The mean of my four numbers is 20."

1. What is the greatest number possible for one of Nathan's numbers? _____

2. How did you figure it out? _____

_____

_____

_____

Describe Data

What Do You Mean?

Eartha

"I am thinking of four counting numbers that are 100 or less.
All of the numbers are different.
The mean of my five numbers is 82."

1. What is the least number possible for one of Eartha's numbers? _____

2. How did you figure it out? _____
_____
_____
_____

What Do You Mean?

"I am thinking of five counting numbers that are 100 or less. All of the numbers are different. The mean of my five numbers is 16."

Linda

1. What is the greatest number possible for one of Linda's numbers? _____

2. How did you figure it out? _____

What Do You Mean?

> I am thinking of five counting numbers that are 80 or less.
> All of the numbers are different.
> The mean of my five numbers is 73.

Marco

1. What is the least number possible for one of Marco's numbers? _____

2. How did you figure it out? _____

What Do You Mean?

I am thinking of six counting numbers that are 50 or less.
All of the numbers are different.
The mean of my six numbers is 42.

Yoshi

1. What is the least number possible for one of Yoshi's numbers? _____

2. How did you figure it out? _____

Solutions

What Do You Mean? 1

1. 39
2. Possible answer: Since the mean of the six numbers is 9, their sum is 6 × 9, or 54. To get the greatest number, first find the sum of the five least counting numbers and then subtract that sum from 54: 1 + 2 + 3 + 4 + 5 = 15, and 54 − 15 = 39. So 39 is the greatest possible number.

What Do You Mean? 2

1. 74
2. Possible answer: Since the mean of the four numbers is 20, their sum is 4 × 20, or 80. To get the greatest number, first find the sum of the three least counting numbers and then subtract that sum from 80: 1 + 2 + 3 = 6, and 80 − 6 = 74. So 74 is the greatest possible number.

What Do You Mean? 3

1. 16
2. Possible answer: Since the mean of the five numbers is 82, their sum is 5 × 82, or 410. To get the least number, first find the sum of the four greatest counting numbers that are 100 or less and then subtract that sum from 410: 100 + 99 + 98 + 97 = 394, and 410 − 394 = 16. So 16 is the least number possible.

What Do You Mean? 4

1. 70
2. Possible answer: Since the mean of the five numbers is 16, their sum is 5 × 16, or 80. To get the greatest number, first find the sum of the four least counting numbers and then subtract that sum from 80: 1 + 2 + 3 + 4 = 10, and 80 − 10 = 70. So 70 is the greatest possible number.

What Do You Mean? 5

1. 51
2. Possible answer: Since the mean of the five numbers is 73, their sum is 5 × 73, or 365. To get the least number, first find the sum of the four greatest counting numbers that are 80 or less and then subtract that sum from 365: 80 + 79 + 78 + 77 = 314, and 365 − 314 = 51. So 51 is the least possible number.

What Do You Mean? 6

1. 12
2. Possible answer: Since the mean of the six numbers is 42, their sum is 6 × 42, or 252. To get the least number, first find the sum of the five greatest counting numbers that are 50 or less and then subtract that sum from 252: 50 + 49 + 48 + 47 + 46 = 240, and 252 − 240 = 12. So 12 is the least possible number.

Number Switch

Goals
- Identify the mean of a set of numbers.
- Understand the relationship between sets of numbers with the same mean and the mean of the combined sets.
- Use logical reasoning to solve problems.

Notes Prior to doing this problem set, review how to compute the mean of a set of numbers. Tell students that all means in this set are whole numbers.

Solutions to all problems in this set appear on page 55.

Number Switch 1

Questions to Ask
- After the switch, suppose the mean of each group was 10. What would the sum of the numbers be in Group A? (40) How do you know? (There are four numbers in Group A, and 4 × 10 = 40.)
- What would the sum of the numbers be in Group B? (60) How do you know? (There are six numbers in Group B, and 6 × 10 = 60.)
- What would the sum of the numbers be in Group A and Group B? (100)
- What would the mean of the numbers be in Group A and Group B? (10)

Solutions
1. Switch 4 (or 6) from Group A with 6 (or 8) from Group B.
2. 8
3. Possible answer: Since the ten numbers in both groups total 80, their mean is 80 ÷ 10, or 8. So the mean of each group must be 8. The sum of the 4 numbers in Group A must be 4 × 8, or 32, and the sum of the 6 numbers in Group B must be 6 × 8, or 48. Group A needs 2 more, and Group B needs 2 less.

GROUNDWORKS

Reasoning with Data and Probability

Number Switch 1

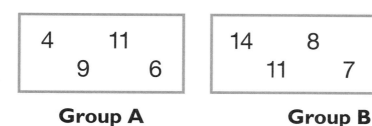

Group A   **Group B**

Switch a number from one group with a number from the other group so that both groups have the same mean.

1. Switch _____ from Group _____ with _____ from Group _____.

2. The mean of the numbers in each group is _____.

3. How did you figure out the numbers to be switched? _____

_____

_____

Number Switch 2

Describe Data

Group A: 7 5 3 6

Group B: 8 4 4 3 5

Switch a number from one group with a number from the other group so that both groups have the same mean.

1. Switch _____ from Group _____ with _____ from Group _____.

2. The mean of the numbers in each group is _____.

3. How did you figure out the numbers to be switched? _____

   _____

   _____

   _____

Number Switch ⟨3⟩

```
  ┌─────────────┐   ┌───────────────────┐
  │  7      9   │   │  6    11      4   │
  │     8       │   │     5      6      │
  └─────────────┘   └───────────────────┘
     Group A              Group B
```

Switch a number from one group with a number from the other group so that both groups have the same mean.

1. Switch _____ from Group _____ with _____ from Group _____.

2. The mean of the numbers in each group is _____.

3. How did you figure out the numbers to be switched? _____

   _____

   _____

Number Switch ◆4

Group A: 16 11 7 7

Group B: 12 8 6 11 7 5

Switch a number from one group with a number from the other group so that both groups have the same mean.

1. Switch _____ from Group _____ with _____ from Group _____.

2. The mean of the numbers in each group is _____.

3. How did you figure out the numbers to be switched? _____

Number Switch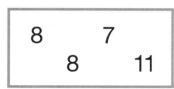

Group A	Group B	Group C
8 7 8 11	9 8 5 7 11	13 0 7 10 9 8 7

Switch a number from one group with a number from another group so that all three groups have the same mean.

1. Switch _____ from Group _____ with _____ from Group _____.

2. The mean of the numbers in each group is _____.

3. How did you figure out the numbers to be switched? _____

   _____

   _____

Number Switch ◆6◆

Group A
10 6
  6 3

Group B
13 9 1
  7 5 8

Group C
4 11 8
  8 12
6 4   5

Switch a number from one group with a number from another group. Then switch a number from one group with a number from another group. Make both switches so that all three groups have the same mean.

1. Switch _____ from Group _____ with _____ from Group _____. Then switch _____ from Group _____ with _____ from Group _____.

2. The mean of the numbers in each group is _____.

3. How did you figure out the numbers to be switched? _____

_____

_____

Describe Data

Solutions

Number Switch 1

1. Switch 4 (or 6) from Group A with 6 (or 8) from Group B.
2. 8
3. Possible answer: Since the ten numbers in both groups total 80, their mean is 80 ÷ 10, or 8. So the mean of each group must be 8. The sum of the four numbers in Group A must be 4 × 8, or 32, and the sum of the six numbers in Group B must be 6 × 8, or 48. Group A needs 2 more, and Group B needs 2 less.

Number Switch 2

1. Switch 5 (or 6) from Group A with 4 (or 5) from Group B.
2. 5
3. Possible answer: Since the nine numbers in both groups total 45, their mean is 45 ÷ 9, or 5. So the mean of each group must be 5. The sum of the four numbers in Group A must be 4 × 5, or 20, and the sum of the five numbers in Group B must be 5 × 5, or 25. Group A needs 1 less and Group B needs 1 more.

Number Switch 3

1. Switch 7 (or 8 or 9) from Group A with 4 (or 5 or 6) from Group B.
2. 7
3. Possible answer: Since the eight numbers in both groups total 56, their mean is 56 ÷ 8, or 7. So the mean of each group must be 7. The sum of the three numbers in Group A must be 3 × 7, or 21, and the sum of the five numbers in Group B must be 5 × 7, or 35. Group A needs 3 less and Group B needs 3 more.

Number Switch 4

1. Switch 11 (or 16) from Group A with 6 (or 11) from Group B.
2. 9
3. Possible answer: Since the ten numbers in both groups total 90, their mean is 90 ÷ 10, or 9. So the mean of each group must be 9. The sum of the four numbers in Group A must be 4 × 9, or 36, and the sum of the five numbers in Group B must be 5 × 9, or 45. Group A needs 5 less and Group B needs 5 more.

Number Switch 5

1. Switch 11 from Group A with 9 from Group C.
2. 8
3. Possible answer: Since the 16 numbers in all three groups total 128, their mean is 128 ÷ 16, or 8. The sum of the four numbers in Group A must be 4 × 8, or 32, the sum of the five numbers in Group B must be 5 × 8, or 40, and the sum of the seven numbers in Group C must be 7 × 8, or 56. Group A needs 2 less and Group C needs 2 more.

Number Switch 6

1. Switch 6 from Group A with 9 from Group B and 6 from Group B with 8 from Group C, or switch 10 from Group A with 13 from Group B and 9 (or 10) from Group B with 11 (or 12) from Group C, or switch 3 from Group A with 6 from Group C and 5 (or 9 or 13) from Group B with 4 (or 8 or 12) from Group C.
2. 7
3. Possible answer: The mean of all 18 numbers is 7. The sum of the numbers in Group A must be 28, in Group B must be 42, and in Group C must be 56.

Reasoning with Data and Probability

Ways to Count

Side by Side

Goals
- Identify permutations of a set of people in a line.
- Create arrangements that match problem conditions.

Notes
Encourage students to consider the number of choices for each seat rather than listing all of the arrangements. For example, if there are 3 people and 1 person wants to sit on the aisle, then the number of choices are 1 for the aisle seat, 2 choices for the next seat, and 1 choice for the next seat, giving a total number of arrangements of $1 \times 2 \times 1$, or 2. Some students may want to write names on index cards, move the cards into different positions, and record the results.

Solutions to all problems in this set appear on page 63.

Side by Side 1

Questions to Ask
- How many boys are there? (5)
- If Alan sits on the aisle, who can sit next to him? (Steven, Dan, John, or Barry)
- If Alan sits on the aisle and Dan sits next to him, who can sit next to Dan? (Steven, John, or Barry)
- What is the total number of sitting arrangements with Alan on the aisle? (24) How did you figure it out? (In each arrangement, Alan's position is fixed and the positions of the other boys vary. The number of choices is 1 for the aisle seat and then 4, 3, 2, and 1 for each of the next four seats, and $1 \times 4 \times 3 \times 2 \times 1 = 24$.)

Solutions
1. 48
2. Possible answer: Since Alan or John must be in the aisle seat, the number of choices for each seat, from the aisle, is 2, 4, 3, 2, and 1. Then the total number of arrangements is $2 \times 4 \times 3 \times 2 \times 1$, or 48.

56 GROUNDWORKS Reasoning with Data and Probability

Side by Side 1

Alan, Steven, Dan, John, and Barry want to sit side by side at the movies. They want to arrange themselves so that Alan or John always sits in the aisle seat.

1. How many different ways can the friends sit side by side? _____

2. How did you figure it out? _____

Name _____

Ways to Count

Side by Side ⟨2⟩

Janet, Roger, Mandy, and Kevin want to sit side by side at a concert. They want to arrange themselves so that Roger always sits in the aisle seat.

1. How many different ways can the friends sit side by side? _____

2. How did you figure it out? _____

_____

_____

58 GROUNDWORKS   Reasoning with Data and Probability

Side by Side ⟨3⟩

Ann, Kim, Jane, and Barbara want to sit side by side at a play. They want to arrange themselves so that Kim or Ann always sits in the aisle seat.

1. How many different ways can the friends sit side by side? _____

2. How did you figure it out? _____

Side by Side ⟨4⟩

Ways to Count

Kate, Sue, Allison, Tami, and Rosa want to sit side by side at the ballet. They want to arrange themselves so that Kate always sits in the middle seat.

1. How many different ways can the friends sit side by side? _____

2. How did you figure it out? _____

Side by Side 5

Ways to Count

Donna, Aretha, Travis, Jody, and Bob want to sit side by side at a soccer game. They want to arrange themselves so that Donna, Jody, or Bob always sits in the aisle seat.

1. How many different ways can the friends sit side by side? _____

2. How did you figure it out? _____

Side by Side ◇6◇

Andrew, Kelly, Ruben, Tony, and Otis want to sit side by side on a bench to take a picture. They want to arrange themselves so that Andrew always sits at one end of the bench and Ruben sits at the other end.

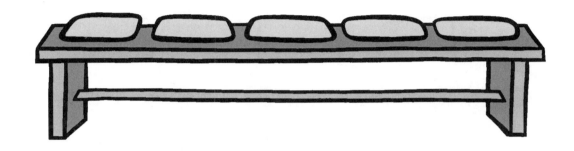

1. How many different ways can the friends sit side by side? _____

2. How did you figure it out? _____

Solutions

Side by Side 1

1. 48
2. Possible answer: Since Alan or John must be in the aisle seat, the number of choices for each seat, from the aisle, is 2, 4, 3, 2, and 1. Then the total number of arrangements is $2 \times 4 \times 3 \times 2 \times 1$, or 48.

Side by Side 2

1. 6
2. Possible answer: Since Roger must be in the aisle seat, the number of choices for each seat, from the aisle, is 1, 3, 2, and 1. Then the total number of arrangements is $1 \times 3 \times 2 \times 1$, or 6.

Side by Side 3

1. 12
2. Possible answer: Since Kim or Ann must be in the aisle seat, the number of choices for each seat, from the aisle, is 2, 3, 2, and 1. Then the total number of arrangements is $2 \times 3 \times 2 \times 1$, or 12.

Side by Side 4

1. 24
2. Possible answer: Since Kate must be in the middle seat, the number of choices for each seat, from the aisle, is 4, 3, 1, 2, and 1. Then the total number of arrangements is $4 \times 3 \times 1 \times 2 \times 1$, or 24.

Side by Side 5

1. 72
2. Possible answer: Since Donna, Jody, or Bob must be in the aisle seat, the number of choices for each seat, from the aisle, is 3, 4, 3, 2, and 1. Then the total number of arrangements is $3 \times 4 \times 3 \times 2 \times 1$, or 72.

Side by Side 6

1. 12
2. Possible answer: Since Andrew and Ruben must be on the ends, the number of choices for each seat, from one end, is 2, 3, 2, 1, and 1. Then the total number of arrangements is $2 \times 3 \times 2 \times 1 \times 1$, or 12.

Ways to Count

Label Makers

Goals
- Apply the Fundamental Counting Principle.
- Find and compare sets of factors.

Notes Suggest that students first identify the number of items possible for a set, then the number of items possible for each slot. They can then apply the Fundamental Counting Principle to determine the number of possible labels. Since the question asks for which is the most, not how many, answers can be determined by comparing the factors of the products without doing the computations.

Solutions to all problems in this set appear on page 71.

Label Makers 1

Questions to Ask
- How many letters are in Label Maker 1 labels? (3)
- How many letters can you choose from? (26)
- If you first choose a letter for the first slot, how many choices are there? (26) If you next choose a letter for the second slot, how many choices are there? (25) How did you decide? (Letters cannot be repeated, so there are 26 − 1, or 25.)
- How many numbers are in Label Maker 1 labels? (3)
- How many choices are there for the first number slot? (10) What are the choices? (any whole number from 0 to 9)
- How many choices are there for the second number slot? (9) The third number slot? (8)

Solutions
1. Label Maker 2
2. Possible answer: Inspection of the factors shows that the factors of Label Maker 2 produce the greatest number.
   LM 1: $26 \times 25 \times 24 \times 10 \times 9 \times 8$
   LM 2: $26 \times 25 \times 24 \times 23 \times 10 \times 9$
   LM 3: $10 \times 10 \times 10 \times 10 \times 10 \times 10$

Label Makers ①

Ways to Count

A label consists of six characters.

☐ — — — — — —

Three different label makers are available.

Label Maker 1
Labels have 3 letters followed by 3 numbers.
Letters and numbers cannot be repeated.

Label Maker 2
Labels have 4 letters followed by 2 numbers.
Letters and numbers cannot be repeated.

Label Maker 3
Labels have 6 numbers.
Numbers can be repeated.

1. Which label maker produces the greatest number of different labels? _____

2. How did you figure it out? _____

_____

_____

_____

Name _____

Ways to Count

Label Makers ⟨2⟩

A label consists of five characters.

[_ _ _ _ _]

Two different label makers are available.

Label Maker 1
Labels have whole numbers.
Numbers can be repeated.

Label Maker 2
Labels have whole numbers.
Numbers cannot be repeated.

1. Which label maker produces the greater number of different labels? _____

2. How did you figure it out? _____

_____

_____

_____

Name _____

Ways to Count

Label Makers ⟨3⟩

A label consists of five characters.

```
  __  __  __  __  __
```

Three different label makers are available.

Label Maker 1
Labels have 3 letters followed by 2 numbers.
Letters and numbers can be repeated.

Label Maker 2
Labels have 2 letters followed by 3 numbers.
Letters and numbers can be repeated.

Label Maker 3
Labels have 5 numbers.
Numbers can be repeated.

1. Which label maker produces the greatest number of different labels? _____

2. How did you figure it out? _____

_____

_____

Reasoning with Data and Probability

GROUNDWORKS ⟨67⟩

Label Makers 4

A label consists of five numbers.

_ _ _ _ _

Three different label makers are available.

Label Maker 1
Labels have only odd numbers.
Numbers can be repeated.

Label Maker 2
Labels have only even numbers.
The number in the first slot cannot be zero.
Numbers can be repeated.

Label Maker 3
Labels have only 0, 1, 2, 3, or 4.
The number in the first slot cannot be zero.
Numbers can be repeated.

1. Which label maker produces the greatest number of different labels? _____

2. How did you figure it out? _____

Name _____

Ways to Count

Label Makers ⟨5⟩

A label consists of six numbers.

[__ __ __ __ __ __]

Three different label makers are available.

Label Maker 1
Labels have whole numbers 3 or less.
The number in the first slot cannot be zero.
Numbers can be repeated.

Label Maker 2
Labels have whole numbers 5 or less.
The number in the first slot cannot be zero.
Numbers cannot be repeated.

Label Maker 3
Labels have only even numbers.
The number in the first slot cannot be zero.
Numbers can be repeated.

1. Which label maker produces the greatest number of different labels? _____

2. How did you figure it out? _____
   _____
   _____
   _____

Reasoning with Data and Probability

GROUNDWORKS ⟨69⟩

Ways to Count

Label Makers 6

A label consists of seven numbers.

[__ __ __ __ __ __ __]

Three different label makers are available.

Label Maker 1
Labels have whole numbers.
The number in the first slot cannot be zero.
Numbers can be repeated.

Label Maker 2
Labels have whole numbers.
A star can be used as a number.
The number in the first slot cannot be zero.
Numbers can be repeated.

Label Maker 3
Labels have whole numbers.
The number in the first slot cannot be zero.
The number in the second slot must be zero.
Numbers can be repeated.

1. Which label maker produces the greatest number of different labels? _____

2. How did you figure it out? _____

_____

_____

_____

GROUNDWORKS   Reasoning with Data and Probability

Solutions

Label Makers 1

1. Label Maker 2
2. Possible answer: Inspection of the factors shows that the factors of Label Maker 2 produce the greatest number.
   LM 1: $26 \times 25 \times 24 \times 10 \times 9 \times 8$
   LM 2: $26 \times 25 \times 24 \times 23 \times 10 \times 9$
   LM 3: $10 \times 10 \times 10 \times 10 \times 10 \times 10$

Label Makers 2

1. Label Maker 1
2. Possible answer: Inspection of the factors shows that the factors of Label Maker 1 produce the greater number.
   LM 1: $10 \times 10 \times 10 \times 10 \times 10$
   LM 2: $10 \times 9 \times 8 \times 7 \times 6$

Label Makers 3

1. Label Maker 1
2. Possible answer: Inspection of the factors shows that the factors of Label Maker 1 produce the greatest number.
   LM 1: $26 \times 26 \times 26 \times 10 \times 10$
   LM 2: $26 \times 26 \times 10 \times 10 \times 10$
   LM 3: $10 \times 10 \times 10 \times 10 \times 10$

Label Makers 4

1. Label Maker 1
2. Possible answer: Inspection of the factors shows that the factors of Label Maker 1 produce the greatest number.
   LM 1: $5 \times 5 \times 5 \times 5 \times 5$
   LM 2: $4 \times 5 \times 5 \times 5 \times 5$
   LM 3: $4 \times 5 \times 5 \times 5 \times 5$

Label Makers 5

1. Label Maker 3
2. Possible answer: Inspection of the factors shows that the factors of Label Maker 3 produce the greatest number.
   LM 1: $3 \times 4 \times 4 \times 4 \times 4 \times 4$
   LM 2: $5 \times 5 \times 4 \times 3 \times 2 \times 1$
   LM 3: $4 \times 5 \times 5 \times 5 \times 5 \times 5$

Label Makers 6

1. Label Maker 2
2. Possible answer: Inspection of the factors shows that the factors of Label Maker 2 produce the greatest number.
   LM 1: $9 \times 10 \times 10 \times 10 \times 10 \times 10 \times 10$
   LM 2: $10 \times 11 \times 11 \times 11 \times 11 \times 11 \times 11$
   LM 3: $9 \times 1 \times 10 \times 10 \times 10 \times 10 \times 10$

Probability

Blocks in the Box

Goals
- Understand probability as a ratio of parts to whole.
- Compute the number of items in a set using a given probability.
- Use logical reasoning to solve problems.

Notes
Prior to doing this problem set, review the relationship between a probability ratio and the number of objects in a set. For example, if the probability of picking a yellow pencil out of a bag of 16 pencils is $\frac{1}{2}$, then there are $\frac{1}{2}$ of 16, or 8, yellow pencils in the bag. If there are 4 green pencils in that bag, then the probability of picking a green pencil is $\frac{4}{16}$, or $\frac{1}{4}$. If the remaining pencils in that bag are purple, then there are 16 − 8 − 4, or 4, purple pencils, and the probability of picking a purple pencil is $\frac{4}{16}$, or $\frac{1}{4}$.

Solutions to all problems in this set appear on page 79.

Blocks in the Box 1

Questions to Ask
- What is the probability of picking a pyramid? ($\frac{1}{2}$)
- If there are 12 blocks in the box, how many blocks are pyramids? (6) How do you know? ($\frac{1}{2}$ of 12 is 6)
- What is the probability of picking a rectangular prism? ($\frac{1}{3}$)
- If there are 12 blocks in the box, how many blocks are rectangular prisms? (4) How do you know? ($\frac{1}{3}$ of 12 is 4)
- What part of the set of blocks are spheres? ($\frac{1}{6}$) How do you know? (Since the fractional parts must add to 1, the spheres are $1 - \frac{1}{2} - \frac{1}{3}$, or $\frac{1}{6}$, of the set of blocks.)

Solutions
1. 42
2. 21
3. 14
4. Possible answer: Since the fractional parts must add to 1, the spheres are $1 - \frac{1}{2} - \frac{1}{3}$, or $\frac{1}{6}$, of the set of blocks. If $\frac{1}{6}$ of the set is 7 blocks, then the entire set has 42 blocks.

Name _____

Probability

Blocks in the Box ⟨1⟩

There are pyramids, rectangular prisms, and spheres in the box. Use the clues to figure out how many blocks are in the box.

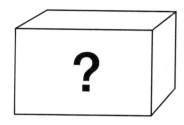

Clues

1. The probability of picking a pyramid is $\frac{1}{2}$.
2. The probability of picking a rectangular prism is $\frac{1}{3}$.
3. There are 7 spheres in the box.

1. How many blocks are in the box? _____

2. How many pyramids are in the box? _____

3. How many rectangular prisms are in the box? _____

4. How did you figure out the answer to Problem 1? _____

_____

_____

_____

_____

Reasoning with Data and Probability

GROUNDWORKS ⟨73⟩

Blocks in the Box 2

There are cylinders, rectangular prisms, and spheres in the box. Use the clues to figure out how many blocks are in the box.

Clues

1. The probability of picking a cylinder is $\frac{1}{4}$.
2. The probability of picking a rectangular prism is $\frac{1}{8}$.
3. There are 5 spheres in the box.

1. How many blocks are in the box? _____

2. How many cylinders are in the box? _____

3. How many rectangular prisms are in the box? _____

4. How did you figure out the answer to Problem 1? _____

Blocks in the Box ⟨3⟩

There are triangular prisms, spheres, and cones in the box. Use the clues to figure out how many blocks are in the box.

Clues

1. The probability of picking a triangular prism is $\frac{5}{8}$.
2. The probability of picking a sphere is $\frac{1}{6}$.
3. There are 10 cones in the box.

1. How many blocks are in the box? _____

2. How many triangular prisms are in the box? _____

3. How many spheres are in the box? _____

4. How did you figure out the answer to Problem 1? _____

_____

_____

_____

Blocks in the Box ⟨4⟩

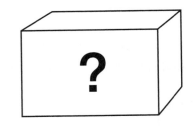

There are tetrahedrons, cubes, and spheres in the box.
Use the clues to figure out how many blocks are in the box.

Clues

1 The probability of picking a tetrahedron is $\frac{3}{10}$.

2 The probability of picking a cube is $\frac{5}{14}$.

3 There are 24 spheres in the box.

1. How many blocks are in the box? _____

2. How many tetrahedrons are in the box? _____

3. How many cubes are in the box? _____

4. How did you figure out the answer to Problem 1? _____

_____

_____

_____

_____

Probability

Blocks in the Box ⟨5⟩

There are cubes, cylinders, triangular prisms, and spheres in the box. Use the clues to figure out how many blocks are in the box.

Clues

1. The probability of picking a cube is $\frac{1}{3}$.
2. The probability of picking a cylinder is $\frac{3}{10}$.
3. The probability of picking a triangular prism is $\frac{1}{5}$.
4. There are 10 spheres in the box.

1. How many blocks are in the box? _____

2. How many cubes are in the box? _____

3. How many cylinders are in the box? _____

4. How many triangular prisms are in the box? _____

5. How did you figure out the answer to Problem 1? _____

_____

_____

_____

Reasoning with Data and Probability

GROUNDWORKS 77

Name _____

Blocks in the Box ⟨6⟩

There are pyramids, rectangular prisms, spheres, and truncated tetrahedrons in the box. Use the clues to figure out how many blocks are in the box.

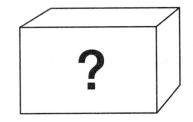

Clues

① The probability of picking a pyramid is $\frac{2}{9}$.

② The probability of picking a rectangular prism is $\frac{7}{30}$.

③ The probability of picking a sphere is $\frac{4}{15}$.

④ There are 50 truncated tetrahedrons in the box.

1. How many blocks are in the box? _____

2. How many pyramids are in the box? _____

3. How many rectangular prisms are in the box? _____

4. How many spheres are in the box? _____

5. How did you figure out the answer to Problem 1? _____

_____

_____

_____

Solutions

Blocks in the Box 1

1. 42
2. 21
3. 14
4. Possible answer: Since the fractional parts must add to 1, the spheres are $1 - \frac{1}{2} - \frac{1}{3}$, or $\frac{1}{6}$, of the set of blocks. If $\frac{1}{6}$ of the set is 7 blocks, then the entire set has 42 blocks.

Blocks in the Box 2

1. 8
2. 2
3. 1
4. Possible answer: Since the fractional parts must add to 1, the spheres are $1 - \frac{1}{4} - \frac{1}{8}$, or $\frac{5}{8}$, of the set of blocks. If $\frac{5}{8}$ of the set is 5 blocks, then the entire set has 8 blocks.

Blocks in the Box 3

1. 48
2. 30
3. 8
4. Possible answer: Since the fractional parts must add to 1, the cones are $1 - \frac{5}{8} - \frac{1}{6}$, or $\frac{5}{24}$, of the set of blocks. If $\frac{5}{24}$ of the set is 10 blocks, then the entire set has 48 blocks.

Blocks in the Box 4

1. 70
2. 21
3. 25
4. Possible answer: Since the fractional parts must add to 1, the spheres are $1 - \frac{3}{10} - \frac{5}{14}$, or $\frac{24}{70}$, or $\frac{12}{35}$, of the set of blocks. If $\frac{12}{35}$, or $\frac{24}{70}$, of the set is 24 blocks, then the entire set has 70 blocks.

Blocks in the Box 5

1. 60
2. 20
3. 18
4. 12
5. Possible answer: Since the fractional parts must add to 1, the spheres are $1 - \frac{1}{3} - \frac{3}{10} - \frac{1}{5}$, or $\frac{5}{30}$, or $\frac{1}{6}$, of the set of blocks. If $\frac{1}{6}$ of the set is 10 blocks, then the entire set has 60 blocks.

Blocks in the Box 6

1. 180
2. 40
3. 42
4. 48
5. Possible answer: Since the fractional parts must add to 1, the truncated tetrahedrons are $1 - \frac{2}{9} - \frac{7}{30} - \frac{4}{15}$, or $\frac{25}{90}$, or $\frac{5}{18}$, of the set of blocks. If $\frac{5}{18}$ of the set is 50 blocks, then the entire set has 180 blocks.

Probability

Marble Odds

Goals
- Recognize the relationship between odds and probability.
- Given the odds, compute the probability of an event.

Notes

Prior to doing this problem set, show students 2 red crayons and 1 black crayon. Point out that probability is *part to whole*; since 1 out of 3 crayons, or $\frac{1}{3}$ of the crayons are black, the probability of picking a black crayon is $\frac{1}{3}$. Explain that, by contrast, odds are *part to part*; the odds in favor of picking a black crayon are 1 (black) to 2 (not black).

Solutions to all problems in this set appear on page 87.

Marble Odds I

Questions to Ask
- What is in the bag? (red and blue marbles)
- What are the odds in favor of picking a red marble? (2 to 3) What does 2 to 3 mean? (Red marbles are 2 parts of the bag and blue marbles are 3 parts of the bag.)
- How many parts are in the entire bag? (5) How do you know? (2 parts red + 3 parts blue = 5 parts total)
- What is the probability of picking a red marble? (2 out of 5, or $\frac{2}{5}$)
- If there are 4 red marbles in the bag, how many marbles must be in the bag? (10) How do you know? (Since the probability of picking a red marble is $\frac{2}{5}$, and there are 2 × 2, or 4, red marbles, there are 2 × 5, or 10, marbles in the bag.

Solutions

1. $\frac{3}{5}$
2. Possible answer: The odds mean that there are 2 red parts for every 3 blue parts, and the entire bag has 5 parts. The probability of picking a blue marble is 3 parts out of 5, or $\frac{3}{5}$.
3. 25
4. Possible answer: Since the probability of picking a red marble is $\frac{2}{5}$, and there are 5 × 2, or 10, red marbles in the bag, there are 5 × 5, or 25, marbles in the bag.

GROUNDWORKS

Reasoning with Data and Probability

Marble Odds 1

There are 10 red marbles and some blue marbles in the bag. The odds in favor of picking a red marble are 2 to 3.

1. What is the probability of picking a blue marble? _____

2. How do you know? _____

_____

_____

3. How many marbles are in the bag? _____

4. How do you know? _____

_____

_____

Name _____

Probability

Marble Odds ⟨2⟩

There are 4 green marbles and some yellow marbles in the bag. The odds in favor of picking a green marble are 1 to 2.

1. What is the probability of picking a yellow marble? _____

2. How do you know? _____
   _____
   _____
   _____

3. How many marbles are in the bag? _____

4. How do you know? _____
   _____
   _____
   _____

GROUNDWORKS   Reasoning with Data and Probability

Marble Odds ⟨3⟩

There are 6 blue marbles and some green marbles in the bag. The odds in favor of picking a green marble are 1 to 3.

1. What is the probability of picking a blue marble? _____

2. How do you know? _____
   _____
   _____
   _____

3. How many marbles are in the bag? _____

4. How do you know? _____
   _____
   _____
   _____

Marble Odds ⟨4⟩

There are 6 blue marbles and some yellow marbles in the bag. The odds in favor of picking a blue marble are 3 to 4.

1. What is the probability of picking a blue marble? _____

2. How do you know? _____
   _____
   _____
   _____

3. How many marbles are in the bag? _____

4. How do you know? _____
   _____
   _____
   _____

Name _____

Probability

Marble Odds ◇5◇

There are 4 blue marbles, some yellow marbles, and some green marbles in the bag. The odds in favor of picking a blue marble are 2 to 5.

1. What is the probability of picking a yellow or a green marble?
   _____

2. How do you know? _____
   _____
   _____
   _____

3. How many marbles are in the bag? _____

4. How do you know? _____
   _____
   _____
   _____

Marble Odds ‹6›

There are 15 blue and green marbles and some red marbles in the bag. The odds in favor of picking a red marble are 3 to 5.

1. What is the probability of picking a red marble? _____

2. How do you know? _____

_____

_____

_____

3. How many marbles are in the bag? _____

4. How do you know? _____

_____

_____

_____

Solutions

Marble Odds 1

1. $\frac{3}{5}$
2. Possible answer: The odds mean that there are 2 red parts for every 3 blue parts, and the entire bag has 5 parts. The probability of picking a blue marble is 3 parts out of 5, or $\frac{3}{5}$.
3. 25
4. Possible answer: Since the probability of picking a red marble is $\frac{2}{5}$, and there are 5 × 2, or 10, red marbles in the bag, there are 5 × 5, or 25, marbles in the bag.

Marble Odds 2

1. $\frac{2}{3}$
2. Possible answer: The odds mean that there is 1 green part for every 2 yellow parts, and the entire bag has 3 parts. The probability of picking a yellow marble is 2 parts out of 3, or $\frac{2}{3}$.
3. 12
4. Possible answer: Since the probability of picking a green marble is $\frac{1}{3}$, and there are 4 × 1, or 4, green marbles in the bag, there are 4 × 3, or 12, marbles in the bag.

Marble Odds 3

1. $\frac{3}{4}$
2. Possible answer: The odds mean that there is 1 green part for every 3 blue parts, and the entire bag has 4 parts. The probability of picking a blue marble is 3 parts out of 4, or $\frac{3}{4}$.
3. 8
4. Possible answer: Since the probability of picking a blue marble is $\frac{3}{4}$, and there are 2 × 3, or 6, blue marbles in the bag, there are 2 × 4, or 8, marbles in the bag.

Marble Odds 4

1. $\frac{3}{7}$
2. Possible answer: The odds mean that there are 3 blue parts for every 4 yellow parts, and the entire bag has 7 parts. The probability of picking a blue marble is 3 parts out of 7, or $\frac{3}{7}$.
3. 14
4. Possible answer: Since the probability of picking a blue marble is $\frac{3}{7}$, and there are 2 × 3, or 6, blue marbles in the bag, there are 2 × 7, or 14, marbles in the bag.

Marble Odds 5

1. $\frac{5}{7}$
2. Possible answer: The odds mean that there are 2 blue parts for every 5 non-blue (yellow or green) parts, and the entire bag has 7 parts. The probability of picking a yellow or a green marble is 5 parts out of 7, or $\frac{5}{7}$.
3. 14
4. Possible answer: Since the probability of picking a blue marble is $\frac{2}{7}$, and there are 2 × 2, or 4, blue marbles in the bag, there are 2 × 7, or 14, marbles in the bag.

Marble Odds 6

1. $\frac{3}{8}$
2. Possible answer: The odds mean that there are 3 red parts for every 5 non-red (blue or green) parts, and the entire bag has 8 parts. The probability of picking a red marble is 3 parts out of 8, or $\frac{3}{8}$.
3. 24
4. Possible answer: Since the probability of picking a blue or green marble is $\frac{5}{8}$, and there are 3 × 5, or 15, blue and green marbles in the bag, there are 3 × 8, or 24, marbles in the bag.

Probability

Mystery Number

Goals
- Compute probabilities.
- Apply the rules of divisibility.
- Use logical reasoning to solve problems.

Notes
Suggest to students that they begin work on each problem by using one or more of the clues to generate a list of candidates for the mystery number. They can then apply the other clues to eliminate candidates until they arrive at one number. Remind students that numbers between two numbers do not include those numbers; for example, if a number is between 200 and 300, it must be greater than 200 and less than 300.

Solutions to all problems in this set appear on page 95.

Mystery Number 1

Questions to Ask
- What is a palindrome? (a word or number that reads the same forward and backward)
- What is the least palindrome between 400 and 600? (404) What is the greatest? (595)
- How can you tell if a number is divisible by 5 without doing the computation? (If the ones digit of the number is 0 or 5, the number is divisible by 5.)
- How can you tell if a number is divisible by 3 without doing the computation? (If the sum of its digits is a multiple of 3, the number is divisible by 3.)

Solutions
1. 585
2. $\frac{3}{199}$

Probability

Mystery Number 1

What is the mystery number?

Clues

1. The mystery number is a whole number between 400 and 600.
2. It is a palindrome.
3. It is divisible by 15.
4. Its tens digit is greater than its hundreds digit.

1. The mystery number is _____.

2. If you randomly chose a whole number between 400 and 600, what is the probability that you would get a palindrome that is divisible by 15?

Mystery Number 2

Clues

1. The mystery number is a whole number between 200 and 300.
2. All of its digits are different even numbers greater than zero.
3. It is divisible by each of its digits.
4. Its tens digit is greater than its ones digits.

1. The mystery number is _____.

2. If you randomly chose a whole number between 200 and 300, what is the probability that you would get the mystery number?

_____

Mystery Number 3

What is the mystery number?

Clues

1. The mystery number is an odd three-digit number.
2. All of its digits are different.
3. Its tens digit is twice its ones digit.
4. Its hundreds digit is a square number.
5. The sum of its digits is 13.

1. The mystery number is _____.

2. If you randomly chose a three-digit number, what is the probability that you would get the mystery number?

_____

Mystery Number 4

What is the mystery number?

Clues

1. The mystery number is a multiple of 10 between 100 and 1,000.
2. Its hundreds digit is twice its tens digit.
3. It is divisible by 4.
4. Its tens digit is not 2.

1. The mystery number is _____.

2. If you randomly chose a whole number between 100 and 1,000, what is the probability that you would get the mystery number?

_____

Mystery Number 5

What is the mystery number?

Clues

1. The mystery number is a three-digit number.
2. Each of its digits is a different square number.
3. The product of its hundreds digit and ones digit is the square of 6.
4. Its ones digit is greater than its hundreds digit.

1. The mystery number is _____.

2. If you randomly chose a three-digit number, what is the probability that you would get a number in which each of its digits was a different square number?

   _____

3. If you randomly chose a three-digit number, what is the probability that you would get a number in which each of its digits was a different square number and the product of its hundreds digit and ones digit was the square of 6?

   _____

Name _____

Probability

Mystery Number 6

What is the mystery number?

Clues

1. The mystery number is a three-digit square number less than 400.
2. It is divisible by 9.
3. It is a multiple of 4.
4. Two of its digits are the same.

1. The mystery number is _____.

2. If you randomly chose a three-digit number less than 400, what is the probability that you would get a square number?

   _____

3. If you randomly choose a three-digit number less than 400, what is the probability that you would get a square number that was divisible by 9?

   _____

Solutions

Mystery Number 1

1. 585
2. $\frac{3}{199}$

Mystery Number 2

1. 264
2. $\frac{1}{99}$

Mystery Number 3

1. 463
2. $\frac{1}{900}$

Mystery Number 4

1. 840
2. $\frac{1}{899}$

Mystery Number 5

1. 419
2. $\frac{6}{900}$, or $\frac{1}{150}$
3. $\frac{2}{900}$, or $\frac{1}{450}$

Mystery Number 6

1. 144
2. $\frac{10}{300}$, or $\frac{1}{30}$
3. $\frac{3}{300}$, or $\frac{1}{100}$

Certificate of Excellence
in Data and Probability

This is to certify that

_____

has satisfactorily completed all the problems for the big idea

_____

and is considered to be an expert.

Date _____   School _____

Grade _____   Teacher _____